OFF GRID SOLAR POWER BIBLE:

A Step-by-Step Guide to Building Your Own Off-Grid Solar Power System and Go from Zero know-how to Zero Electricity Bills

BY

LESTER VIENS

First published by
Lester Viens

TABLE OF CONTENTS

INTRODUCTION

In a world increasingly driven by the need for sustainable and environmentally friendly energy sources, off-grid solar power has emerged as a beacon of hope for those seeking a more self-reliant, eco-conscious, and resilient way of life. The Off-Grid Solar Power Bible is a comprehensive guide that illuminates the path to harnessing the boundless energy of the sun to power our homes, businesses, and communities independently of traditional electrical grids. As we stand on the precipice of a renewable energy revolution, this Bible serves as a beacon of knowledge, offering guidance, insights, and practical advice to empower individuals, families, and communities to break free from the shackles of conventional power systems and embrace the limitless potential of off-grid solar solutions.

The growing concern over climate change, dwindling fossil fuel reserves, and the increasing strain on centralized energy infrastructure has compelled us to seek alternative energy sources. Solar power, with its ability to provide clean and abundant energy, has emerged as a frontrunner in the race to transition to a more sustainable future. However, the true potential of solar power lies not only in its capacity to reduce our carbon footprint but also in its ability to grant us energy independence. Off-grid solar systems, which are free from the constraints of utility companies and transmission lines, have revolutionized the way we generate, store, and utilize electricity.

The Off-Grid Solar Power Bible will be your guiding light on this transformative journey. Whether you're an environmentally conscious individual yearning for an independent lifestyle, a homesteader seeking to create a self-sufficient haven, a community looking to develop microgrids, or a business owner seeking energy security, this comprehensive resource will provide you with the knowledge and skills to make your off-grid solar dreams a reality.

Through a careful exploration of solar technology, energy storage, system design, installation, maintenance, and best practices, this Bible will help you understand the fundamental concepts of off-grid solar power. From the basics of photovoltaics and battery technology to the intricacies of sizing and configuring solar arrays, this guide offers a wealth of information for both beginners and experienced off-grid enthusiasts. The Off-Grid Solar Power Bible empowers you to choose the right equipment, optimize your energy consumption, and navigate the financial aspects of your solar journey, including incentives and return on investment.

Moreover, the Off-Grid Solar Power Bible transcends the technical realm and delves into the broader implications of adopting off-grid solar solutions. It explores the socio-economic and environmental benefits, the potential for energy security in remote areas, and the ways in which off-grid systems can foster resilience in the face of disasters and crises. Furthermore, it addresses the role of off-grid solar power in reducing energy poverty and bridging the global energy divide.

As we embark on this exciting journey into the world of off-grid solar power, the Off-Grid Solar Power Bible stands as a beacon of knowledge and a testament to the incredible potential of harnessing the sun's energy to transform our lives, our communities, and our planet. Whether you're a curious novice or an experienced off-grid pioneer, this comprehensive guide will illuminate your path towards a brighter, cleaner, and more self-reliant future. So, let us embark on this solar-powered odyssey together, for a world powered by the sun is a world full of possibilities.

CHAPTER 1: UNDERSTANDING OFF-GRID SOLAR POWER

The Basics of Solar Energy

Solar power is a clean and sustainable source of energy that has gained immense popularity in recent years. In particular, off-grid solar power systems have become a viable solution for individuals and communities looking to access electricity in remote or underserved areas. This article aims to provide you with a comprehensive understanding of off-grid solar power by first delving into the basics of solar energy.

The Basics of Solar Energy

Solar energy, often referred to as photovoltaic (PV) energy, is harnessed from the sun's radiant light and converted into electricity. Understanding how this process works is crucial to grasp the foundation of off-grid solar power.

Solar Panels: Solar panels, or solar photovoltaic cells, are the primary components of a solar energy system. These panels are made from semiconductor materials, typically silicon, and are designed to capture sunlight. When sunlight hits the panels, it excites electrons in the semiconductor, creating a flow of electricity.

Photovoltaic Effect: The photovoltaic effect, discovered by Alexandre-Edmond Becquerel in 1839, is the phenomenon where certain materials generate an electric current when exposed to light. This effect is the fundamental principle behind solar energy conversion.

Inverters: The electricity generated by solar panels is in the form of direct current (DC), but most household appliances and electronics operate on alternating current (AC). Inverters are used to convert the DC electricity into AC, making it suitable for powering your home.

Solar Charge Controllers: Off-grid solar power systems often include batteries to store excess energy for use when the sun is not shining. Solar charge controllers regulate the charging and discharging of these batteries, preventing overcharging and deep discharging, which can damage the batteries.

Battery Storage: Batteries are an essential component in off-grid solar systems, as they store surplus energy for use during the night or on cloudy days. Common battery types include lead-acid, lithium-ion, and deep-cycle batteries.

Off-Grid vs. Grid-Tied Solar Systems: The key difference between off-grid and grid-tied solar systems is that off-grid systems are independent and not connected to the utility grid. Grid-tied systems, on the other hand, feed excess energy back into the grid and draw power from it when needed. Off-grid systems are ideal for remote areas or for those looking to be self-sufficient.

Energy Efficiency and Conservation: Maximizing energy efficiency and practicing energy conservation is essential in off-grid systems. By using energy-efficient appliances and reducing unnecessary energy consumption, you can make the most of the solar power generated.

Understanding the basics of solar energy is the first step in comprehending the operation of off-grid solar power systems. As the world shifts towards more sustainable and eco-friendly energy sources, off-grid solar power is becoming increasingly attractive for those seeking independence from traditional utility services. In addition to providing electricity in remote areas, off-grid solar power systems are environmentally friendly and can lead to significant cost savings over time. In the next installment of our series on off-grid solar power, we will explore the components and workings of these systems in more detail, shedding light on how you can harness the sun's energy to power your off-grid lifestyle.

Advantages of Off-Grid Solar Power

Off-grid solar power systems have gained popularity in recent years as a sustainable and reliable solution for generating electricity in areas where traditional grid connections are unavailable or unreliable. These systems utilize solar panels, batteries, and inverters to harness energy from the sun and store it for later use. Understanding the advantages of off-grid solar power is crucial for those considering this technology as an alternative source of energy. In this article, we'll explore the numerous benefits of off-grid solar power.

Energy Independence

One of the most significant advantages of off-grid solar power is energy independence.

With an off-grid solar system, you're not reliant on a central grid or utility company for your electricity needs. This independence allows you to generate, store, and use your electricity, reducing your dependence on fossil fuels and providing a reliable source of power, even in remote locations.

Cost Savings

Off-grid solar power systems can lead to substantial cost savings over time. While the initial investment in solar panels, batteries, and inverters may seem high, the long-term operational costs are significantly lower. Once the system is in place, sunlight is free, and you won't be subject to rising utility rates. Plus, some governments offer incentives and tax credits to encourage the adoption of renewable energy, making the upfront costs more affordable.

Environmentally Friendly

Off-grid solar power is an environmentally friendly energy solution. Solar panels produce electricity without emitting greenhouse gases, reducing your carbon footprint and helping combat climate change. By using clean, renewable energy, you're contributing to a more sustainable future.

Reliability

Off-grid solar power systems are known for their reliability. They can provide electricity even in remote or off-the-grid locations where traditional power sources are unavailable. With proper system design and maintenance, off-grid solar power can ensure a continuous and uninterrupted energy supply.

Reduced Noise Pollution

Unlike traditional generators that can be noisy and disruptive, off-grid solar power systems operate silently. This not only creates a more peaceful and comfortable living environment but also reduces noise pollution, making it an ideal solution for residential areas and places where tranquility is valued.

Minimal Maintenance

Off-grid solar power systems require minimal maintenance. Solar panels are durable and can last for 25 years or more, with little to no maintenance needed. Batteries may require occasional checks, but modern battery technology is increasingly reliable and long-lasting.

Scalability

Off-grid solar power systems are highly scalable. You can start with a small system and expand it as your energy needs grow. This flexibility allows you to adapt your system to your changing circumstances, making off-grid solar an ideal solution for both residential and commercial applications.

Off-grid solar power offers a range of advantages that make it an attractive option for those seeking energy independence, cost savings, and environmental benefits. By harnessing the power of the sun, off-grid solar systems provide a reliable and sustainable source of electricity while reducing the carbon footprint. With advancements in technology and government incentives, off-grid solar power has become more accessible and affordable than ever, making it a viable choice for a greener and more self-reliant future.

Is Off-Grid Solar Right for You?

In an era where renewable energy solutions are gaining prominence, off-grid solar power systems have emerged as a promising and sustainable option for both rural and urban residents seeking independence from traditional grid-based electricity. These systems offer an alternative means of generating and storing energy, but the question remains: Is off-grid solar power the right choice for you? To answer that question, it's crucial to understand the workings and benefits of off-grid solar power and consider your unique needs and circumstances.

Understanding Off-Grid Solar Power

Off-grid solar power, also known as standalone or independent solar power, is a decentralized energy solution that allows you to generate, store, and use electricity independently, without relying on a central grid. This technology primarily consists of photovoltaic (PV) panels that convert sunlight into electricity, a battery bank to store excess energy, and an inverter to convert direct current (DC) into alternating current (AC) for your electrical appliances.

Is Off-Grid Solar Right for You?

To determine if off-grid solar power is the right choice for your energy needs, you should consider the following factors:

Location: One of the most critical factors to consider is your location. Off-grid solar power is particularly beneficial in remote areas where grid access is limited or expensive to install. If you live in a rural or off-grid location, off-grid solar could be an excellent option to gain energy independence.

Energy Requirements: Evaluate your energy consumption and requirements. Off-grid solar systems are suitable for those who are willing to adapt their energy usage to match the available solar energy. If you are energy-conscious and can reduce your electricity consumption during cloudy or nighttime periods, off-grid solar may work well for you.

Initial Investment: Off-grid solar systems require an upfront investment in solar panels, batteries, and inverters. While there may be government incentives and rebates available, it's essential to assess whether the initial cost aligns with your budget and long-term savings.

Maintenance and Self-Sufficiency: Off-grid solar systems require more hands-on maintenance and technical knowledge compared to grid-connected systems. You should be prepared to maintain and troubleshoot your system or have access to local experts who can assist you.

Environmental Concerns: Off-grid solar power is a green energy solution, reducing your carbon footprint and reliance on fossil fuels. If environmental sustainability is a priority for you, off-grid solar is a great choice.

Backup Generator: Consider whether you need a backup generator for extended periods of bad weather or increased energy demand. Some off-grid users opt for a generator to provide power during these situations.

Energy Storage Capacity: The size of your battery bank is critical in determining the autonomy of your system. Ensure that your system has enough storage capacity to meet your energy needs during cloudy days or at night.

Planning and Design: Proper planning and system design are essential to ensure your off-grid solar system meets your needs. Consult with a qualified solar installer to assess your location and develop a customized system.

Off-grid solar power offers energy independence, sustainability, and resilience to those willing to embrace the lifestyle it entails. While it may not be the right choice for everyone, it can be a fantastic option for those living in remote areas, environmentally conscious individuals, and those seeking long-term savings on their energy bills.

If you're considering going off-grid with solar power, carefully assess your location, energy requirements, budget, and maintenance capabilities. Seek expert advice to ensure your system is tailored to your unique needs. With the right planning and knowledge, off-grid solar power can provide you with a reliable and sustainable source of energy.

CHAPTER 2: PLANNING YOUR OFF-GRID SOLAR POWER SYSTEM

Assessing Your Energy Needs

One of the most critical steps in setting up an off-grid solar power system is assessing your energy needs. A thorough understanding of your energy requirements will serve as the foundation for designing an efficient and cost-effective solar power system that can meet your needs. In this article, we'll guide you through the process of assessing your energy needs when planning an off-grid solar power system.

Calculate Your Daily Energy Consumption:

Before you can determine the size and capacity of your off-grid solar power system, you need to know how much energy your household or facility consumes on a daily basis. Start by making a list of all the electrical appliances and devices you use, along with their power ratings in watts (W) or kilowatts (kW). Include lighting, refrigeration, heating, cooling, entertainment systems, and any other equipment that consumes electricity. Be as thorough as possible to ensure accurate calculations.

Determine Peak Load:

Next, identify the peak load, which represents the highest simultaneous energy demand in your household or facility. This is typically the sum of all the high-power appliances running at the same time, such as air conditioning, water heaters, or power tools. Peak load is crucial for sizing the inverter and battery bank, as they must handle these spikes in energy demand.

Assess Seasonal Variations:

Consider any seasonal variations in your energy consumption. For example, you may use more energy for heating or cooling during extreme weather conditions. This information will help you size your solar power system to meet your energy needs year-round.

Evaluate Energy Efficiency:

An important aspect of assessing your energy needs is to evaluate your energy efficiency. Are there ways to reduce energy consumption through energy-efficient appliances, insulation, or behavioral changes? Lowering your energy demands can result in a smaller and more cost-effective solar power system.

Account for Standby Loads:

Don't forget to account for standby loads, which are devices that consume energy even when not actively in use. Examples include chargers, standby power for electronics, and appliances on standby mode. These loads can add up over time and should be included in your energy assessment.

Analyze Solar Radiation Data:

To accurately size your off-grid solar power system, you must analyze the solar radiation data specific to your location. The amount of daily sunlight and its seasonal variations will impact the size and orientation of your solar panels. Consider factors like shading, cloud cover, and local climate conditions when analyzing solar radiation data.

Determine Battery Capacity:

Your battery bank size is closely tied to your energy needs, as it stores excess energy generated by your solar panels for use during cloudy days or at night. The capacity of your battery bank should be able to meet your energy requirements during periods of low or no solar generation.

Consult a Solar Professional:

For a comprehensive energy assessment and system design, it is highly recommended to consult a professional solar installer or engineer. They can help you perform a detailed energy audit, taking into account all the variables unique to your location and lifestyle.

Assessing your energy needs is the crucial first step in planning an off-grid solar power system. By calculating your daily energy consumption, determining peak loads, accounting for seasonal variations, and analyzing solar radiation data, you can design a system that meets your energy requirements while maximizing energy efficiency and reducing environmental impact. Consulting with a professional will help ensure that your off-grid solar power system is reliable and tailored to your specific needs.

Site Assessment and Solar Panel Placement

Off-grid solar power systems provide an environmentally friendly and sustainable way to generate electricity in remote locations or areas without access to the grid. To ensure the successful and efficient operation of your off-grid solar power system, careful planning is essential. This article focuses on the initial steps of site assessment and solar panel placement, which are crucial in maximizing energy production and system performance.

Site Assessment:

Before you start purchasing solar panels and other components for your off-grid solar power system, a comprehensive site assessment is imperative. Here's how to conduct one:

Location Selection:

Identify the specific location where you intend to install your solar system. It should receive ample sunlight throughout the year.

Consider factors such as latitude, longitude, and elevation, as they impact the angle and duration of sunlight available.

Sunlight Availability:

Evaluate the solar irradiance in your area, which measures the amount of solar energy that reaches your location. This data is often available through local weather services or online solar calculators.

Consider seasonal variations in sunlight and weather patterns when assessing energy availability.

Shading Analysis:

Identify potential sources of shading, like trees, buildings, or terrain features that could block sunlight.

Monitor the shading patterns throughout the day and year to determine their impact on your system's performance.

Energy Needs:

Calculate your energy consumption needs to determine the size of the solar system required.

Analyze your daily and seasonal electricity usage to ensure your system is appropriately sized.

Solar Panel Placement:

Once you've completed the site assessment, the next step is to plan the placement of your solar panels to maximize energy production:

Solar Panel Orientation:

Position your solar panels to face the equator (south in the Northern Hemisphere, north in the Southern Hemisphere) for optimal sun exposure. The angle of tilt should be equal to your latitude for maximum annual energy production. However, you may need to adjust this angle to optimize performance during specific seasons.

Array Design:

Arrange your solar panels in arrays, considering the available space and the system's voltage requirements.

Ensure adequate spacing between panels to prevent shading and allow for cooling, which improves efficiency.

Tracking Systems:

For improved energy generation, consider installing tracking systems that follow the sun's movement during the day. Single-axis or dual-axis tracking systems can significantly increase energy output

Mounting Options:

Choose between ground-mounted and roof-mounted installations. Ground-mounted systems are easier to install and maintain, but roof-mounted systems can save space and reduce shading issues.

Tilt and Elevation:

Pay attention to the tilt and elevation of your solar panels. Panels should be elevated off the ground to prevent damage and facilitate maintenance.

Safety and Aesthetics:

Ensure that your solar panel placement complies with local regulations and safety standards. Consider the visual impact of your installation on the surrounding environment.

Proper site assessment and solar panel placement are critical steps in planning your off-grid solar power system. By carefully evaluating your location and optimizing your panel placement, you can maximize energy production, enhance system performance, and ultimately achieve energy independence. Whether you're looking to power a remote cabin or reduce your carbon footprint, a well-planned off-grid solar system can provide you with clean and sustainable energy for years to come.

Battery Storage Solutions

As the world transitions towards cleaner and more sustainable energy sources, off-grid solar power systems have gained popularity as an environmentally friendly and cost-effective way to generate electricity. These systems can provide reliable energy even in remote locations or during power outages, making them an attractive option for homeowners, businesses, and remote communities. When designing an off-grid solar power system, one of the key components to consider is battery storage solutions. In this guide, we'll explore the crucial aspects of planning your off-grid solar power system with a focus on battery storage solutions.

Why Battery Storage Matters:

Battery storage is the heart of an off-grid solar power system. While solar panels capture and convert sunlight into electricity, batteries store and provide power when the sun is not shining, such as during the night or on cloudy days. These batteries play a vital role in ensuring a continuous and reliable power supply, making them a critical component for those looking to go off-grid.

Factors to Consider in Battery Selection:

Selecting the right battery for your off-grid solar system is a decision that should not be taken lightly. Here are some key factors to consider:

Battery Type:

Lead-Acid Batteries: These are the traditional and more affordable option. However, they have a shorter lifespan and lower energy density compared to newer technologies.

Lithium-Ion Batteries: Lithium-ion batteries have a longer lifespan, higher energy density, and are more compact. While they tend to be more expensive upfront, their cost-effectiveness over the long term makes them a popular choice for off-grid systems.

Capacity (Ah):

The capacity of a battery is measured in ampere-hours (Ah) and determines how much energy it can store. Calculate your daily energy consumption to determine the battery capacity needed for your system.

Depth of Discharge (DoD):

The DoD indicates how much of a battery's capacity can be used before it should be recharged. A deeper DoD allows you to use more of the battery's capacity, but it can impact its lifespan.

Lifespan:

Batteries have a limited lifespan, typically measured in cycles. Consider how many cycles a battery is rated for and its warranty when making your choice.

Maintenance:

Some batteries require regular maintenance, such as topping up electrolyte levels in lead-acid batteries. Lithium-ion batteries are typically maintenance-free.

Temperature Tolerance:

Batteries perform best within a certain temperature range. Consider the climate of your location and whether your chosen battery can handle temperature extremes.

System Design and Sizing:

Properly sizing your battery bank is crucial for the efficient operation of your off-grid solar system. Oversized or undersized batteries can lead to inefficiencies or system failure. Here are the steps for determining the size of your battery bank:

Calculate Daily Energy Consumption:

Determine how much energy your appliances and devices consume daily. This will be the basis for sizing your battery bank.

Account for Days of Autonomy:

Consider how many days of autonomy you need. This is the number of days your system can operate without recharging the batteries. It depends on your location and how frequently sunlight is available.

Calculate Battery Capacity:

Multiply your daily energy consumption by the number of days of autonomy. Then, divide this by the depth of discharge to find the required battery capacity (Ah).

Choose Battery Voltage:

The voltage of your battery bank will depend on your inverter's voltage rating. Common options are 12V, 24V, or 48V systems.

Planning an off-grid solar power system with battery storage solutions is a complex but rewarding endeavor. By selecting the right batteries and sizing your system correctly, you can enjoy reliable electricity even in remote locations. As technology continues to advance, battery storage options are becoming more efficient and cost-effective, making off-grid living a viable and sustainable choice for a growing number of people. Properly planned battery storage is the key to unlocking the full potential of your off-grid solar system.

CHAPTER 3: SELECTING AND INSTALLING SOLAR COMPONENTS

Choosing the Right Solar Panels

Solar energy is a clean and renewable source of power that is becoming increasingly popular for residential and commercial use. One of the key components of a solar energy system is the solar panels. Choosing the right solar panels is crucial to ensure that your solar energy system operates efficiently and provides you with the maximum return on your investment. In this guide, we will explore the factors to consider when selecting solar panels for your project.

Types of Solar Panels

There are several types of solar panels available on the market, with the most common being monocrystalline, polycrystalline, and thin-film solar panels. Each type has its unique characteristics and is suited for different applications:

Monocrystalline Solar Panels: Monocrystalline panels are known for their high efficiency and sleek black appearance. They are made from single-crystal silicon, making them more efficient in converting sunlight into electricity. These panels are ideal for projects with limited roof space.

Polycrystalline Solar Panels: Polycrystalline panels are less expensive than monocrystalline panels but slightly less efficient. They are made from multiple silicon fragments, giving them a blue-ish appearance. These panels are a cost-effective choice for larger installations.

Thin-Film Solar Panels: Thin-film panels are lightweight and flexible, making them suitable for unconventional installation methods. However, they are less efficient than crystalline panels and often require more space.

Efficiency

Solar panel efficiency is a critical factor to consider. It determines how much sunlight a panel can convert into electricity. The higher the efficiency, the more electricity you can generate from the same amount of sunlight. Efficiency is typically expressed as a percentage and can range from 15% to 22% or more for top-tier panels. While higher efficiency panels tend to cost more, they can be a better long-term investment as they produce more power from the same physical space.

Durability and Warranty

The longevity and reliability of your solar panels are essential. Consider the panel's durability, as it needs to withstand various weather conditions, including rain, wind, hail, and snow. Check for certifications such as IEC, UL, or TÜV, which indicate quality and safety standards.

Additionally, examine the warranty offered by the panel manufacturer. A standard warranty for solar panels typically covers 25 years, but the performance guarantee may vary. Be sure to understand what aspects of the panel's performance the warranty covers and for how long.

Cost

Cost is a significant consideration when selecting solar panels. The initial cost of the panels and installation can vary depending on the type, brand, and efficiency of the panels. It's essential to find a balance between your budget and the long-term energy savings the panels can provide.

Consider the cost of the entire solar energy system, including inverters, racking, and installation, to get a comprehensive picture of your investment. While it may be tempting to opt for the least expensive option, remember that higher-quality, more efficient panels can pay off in the form of lower energy bills and a quicker return on investment.

Aesthetics

The appearance of your solar panels might be important, especially for residential installations. Some homeowners prefer panels with a sleek and uniform look, such as monocrystalline panels, while others may not mind the appearance of polycrystalline or thin-film panels. Your choice will depend on your personal preferences and the aesthetics of your property.

Selecting the right solar panels is a crucial step in building an efficient and effective solar energy system. Consider the type, efficiency, durability, cost, warranty, and aesthetics when making your decision. It's also a good idea to consult with a professional solar installer who can assess your specific needs and help you choose the best panels for your project. With careful consideration and the right components, you can harness the power of the sun to reduce your energy costs and environmental footprint.

Inverters, Charge Controllers, and Wiring

Harnessing the power of the sun through solar energy is not just about installing photovoltaic panels on your roof. A successful solar energy system relies on several essential components, including inverters, charge controllers, and proper wiring. In this guide, we will explore the crucial aspects of selecting and installing these components to ensure the efficient and reliable operation of your solar energy system.

Inverters:

Inverters play a pivotal role in a solar energy system as they convert the direct current (DC) electricity generated by your solar panels into alternating current (AC) electricity, which is suitable for powering your home and feeding excess energy back into the grid. Here's how to choose and install the right inverter for your solar system:

Types of Inverters:

i. **String Inverters:** These are cost-effective and suitable for small to medium-sized systems. They are typically installed near the solar panels.

ii. **Microinverters:** Each panel has its own microinverter, offering maximum efficiency and flexibility but at a higher cost.

iii. **Hybrid Inverters:** These inverters are essential if you plan to incorporate energy storage solutions, like batteries, into your solar system.

Sizing Your Inverter:

Properly size your inverter to match the capacity of your solar array. An oversized inverter can lead to inefficiency, while an undersized inverter can limit the system's performance.

Installation:

Ensure that your inverter is installed in a cool, well-ventilated location to prevent overheating. It should also be protected from the elements.

Charge Controllers:

Charge controllers are essential for ensuring that your batteries are charged efficiently while preventing overcharging and damage to your batteries. Here's what to consider when selecting and installing charge controllers:

Types of Charge Controllers:

i. PWM (Pulse Width Modulation) controllers are cost-effective and suitable for smaller systems.

ii. MPPT (Maximum Power Point Tracking) controllers are more efficient and adapt well to various solar panel configurations.

Sizing Your Charge Controller:

Choose a charge controller that can handle the current from your solar panels and is compatible with your battery bank.

Installation:

Install the charge controller between the solar panels and the battery bank. Ensure proper wiring and follow the manufacturer's guidelines for installation.

Wiring:

Wiring is the backbone of your solar energy system, carrying electricity from the solar panels to the inverter, charge controller, and the rest of your electrical system. Proper wiring is crucial for safety and efficiency. Consider the following when selecting and installing wiring for your solar system:

Cable Type:

Use high-quality, weather-resistant, and UV-resistant cables that can withstand outdoor conditions. Copper wiring is the most common choice for solar applications.

Wire Sizing:

Properly size your wires to minimize energy loss due to resistance. Consult the National Electric Code (NEC) or a professional installer for guidance.

Conduit and Grounding:

Install conduits to protect the wires from physical damage and ensure proper grounding to reduce the risk of electrical hazards.

Selecting and installing the right inverters, charge controllers, and wiring are critical steps in creating a reliable and efficient solar energy system. It's essential to choose components that match your specific requirements, follow safety guidelines, and consider the long-term performance and maintenance of your system. If you're not confident in your abilities, it's advisable to consult with a professional solar installer who can ensure that your components are selected and installed correctly, maximizing the benefits of your solar investment.

Mounting and Installing Solar Panels

Solar panels are a key component of any solar energy system, and their proper mounting and installation are crucial for the system's performance and longevity. Whether you're a homeowner looking to reduce your energy bills or a business owner aiming to go green, selecting and installing solar panels requires careful consideration. In this guide, we will explore the essential factors to consider when mounting and installing solar panels.

Selecting the Right Solar Panels

Before you can install solar panels, you must choose the right type and model. Here are some factors to consider:

Solar Panel Type

There are three main types of solar panels: monocrystalline, polycrystalline, and thin-film. Monocrystalline panels are known for their efficiency and sleek design, while polycrystalline panels are more cost-effective. Thin-film panels are lightweight and versatile but less efficient. Your choice will depend on your budget and available space.

Efficiency

Consider the efficiency rating of the solar panels. Higher efficiency panels will generate more electricity in a smaller space, which can be beneficial if space is limited or aesthetics are a concern.

Warranty

Check the manufacturer's warranty. A longer warranty period indicates the manufacturer's confidence in the durability of their product.

Aesthetics

The appearance of the panels may be important, especially for residential installations. Some homeowners prefer sleek, all-black panels that blend seamlessly with their roof.

Mounting Solar Panels

Mounting solar panels properly is essential to ensure they are secure and receive optimal sun exposure. Here's what you need to know:

Roof or Ground Mounting

Solar panels can be installed on rooftops or as ground-mounted systems. Roof mounting is a popular choice for residential installations, while ground mounting is common for larger systems or when the roof isn't suitable.

Tilt and Orientation

To maximize energy production, solar panels should face true south (in the Northern Hemisphere) and be tilted at an angle equal to the latitude of your location. However, local shading and other factors may require adjustments to the orientation.

Racking System

A racking system is used to attach the panels to the mounting surface. Ensure that the racking system is durable, weather-resistant, and compatible with your chosen solar panels.

Secure Attachment

Properly fasten the solar panels to the mounting structure to prevent wind damage or other environmental factors that may dislodge them.

Wiring and Inverter

Ensure that wiring is neat and well-organized to prevent damage and improve safety. Connect the panels to an inverter, which converts the direct current (DC) generated by the panels into alternating current (AC) usable for your home or business.

Safety Considerations

When installing solar panels, safety should be a top priority. Here are some safety considerations:

Hire Professionals

While some people may attempt DIY solar panel installation, it's often safer and more efficient to hire professionals with experience in the field.

Compliance

Ensure your installation complies with local building codes, electrical regulations, and safety standards.

Roof Inspection

Before installing solar panels on a roof, have it inspected to ensure it can support the additional weight.

Protection from Fire

Install fire-resistant materials under and around the solar panels to reduce the risk of fire.

Grounding

Properly ground the solar system to minimize the risk of electrical hazards.

Maintenance

Once your solar panels are installed, regular maintenance is essential to ensure they operate efficiently over the years. This includes cleaning the panels, checking for loose connections, and monitoring the system's performance.

In conclusion, selecting and installing solar panels is a significant investment that can provide long-term benefits in terms of energy savings and reducing your carbon footprint. By carefully choosing the right panels and following best practices for mounting and installation, you can enjoy the full potential of your solar energy system for many years to come. If you're unsure about any aspect of the process, it's wise to consult with a professional solar installer to ensure a successful and safe installation.

CHAPTER 4: SYSTEM MAINTENANCE AND TROUBLESHOOTING

Maintenance Tips for Optimal Performance

Effective system maintenance is essential to keep your devices, whether they are computers, smartphones, or other electronics, running smoothly and performing at their best. Neglecting maintenance can lead to decreased performance, system errors, and even hardware failure. This article will provide you with valuable maintenance tips to ensure optimal performance for your systems.

Regular Software Updates:

Keeping your operating system and software applications up to date is crucial for system performance. Software updates often include bug fixes, security patches, and performance enhancements that can help your system run more efficiently. Set your system to install updates automatically or check for updates manually on a regular basis.

Clean Your Device:

Dust and debris can accumulate in and around your device, causing it to overheat and slow down. To prevent this, periodically clean your device by using compressed air to remove dust from fans and vents. For laptops, consider cleaning the keyboard and screen as well.

Organize and Delete Unnecessary Files:

Over time, your system can become cluttered with unnecessary files, such as old documents, temporary files, and cached data. Regularly clean up your files and uninstall programs you no longer need to free up storage space and improve system performance.

Backup Your Data:

Regular data backups are essential in case of system failures, crashes, or data loss. Use an external hard drive, cloud storage, or a backup service to ensure your important files and documents are safe and can be restored if needed.

Run Antivirus and Malware Scans:

Protect your system from malware and viruses by regularly running antivirus and anti-malware scans. These scans can identify and remove any malicious software that might be affecting your system's performance.

Manage Startup Programs:

Many applications and services start automatically when you boot up your system. Some of these may not be necessary for everyday use. Review and disable unnecessary startup programs to improve your system's boot time and overall performance.

Monitor System Temperatures:

Overheating can lead to system slowdowns and potential damage to your hardware. Use monitoring software to check your system's temperature, and if it's consistently running hot, consider cleaning the internal components or investing in additional cooling solutions.

Check for Hardware Issues:

If your system experiences frequent crashes, freezes, or other performance issues, it may be related to hardware problems. Run diagnostic tests or seek the help of a professional technician to identify and resolve any hardware issues.

Optimize Your Browser:

If you spend a lot of time browsing the internet, optimize your web browser by clearing the cache and cookies regularly. Consider using browser extensions or add-ons that can enhance performance and security.

Implement a System Restore Point:

Create regular system restore points so that if something goes wrong with your system, you can revert it to a previous state when it was functioning correctly.

Maintaining your system is not only essential for optimal performance but also for extending the lifespan of your devices. By following these maintenance tips, you can ensure that your system runs smoothly, efficiently, and remains free from potential issues that may disrupt your workflow. Regular system maintenance will save you time, money, and frustration in the long run.

Identifying and Fixing Common Issues

System maintenance is an essential part of keeping your computer, network, or any other technological device running smoothly. However, even with the best care, issues can arise from time to time. In this guide, we'll explore the importance of regular system maintenance and provide valuable insights into identifying and fixing common issues that may disrupt your system's functionality.

The Importance of System Maintenance

Regular system maintenance is crucial to ensure the optimal performance and longevity of your devices. Neglecting maintenance can lead to various problems, such as decreased system speed, data loss, and even complete system failure. Here are some key reasons why system maintenance is essential:

Improved Performance: Routine maintenance tasks, such as cleaning up temporary files, defragmenting hard drives, and updating software, can significantly enhance your system's speed and responsiveness.

Data Protection: Maintenance tasks like regular backups and antivirus scans help protect your valuable data from loss or damage due to malware, hardware failures, or other unforeseen issues.

Longevity: Proper maintenance can extend the lifespan of your devices and reduce the need for costly replacements.

Enhanced Security: Keeping your system up to date with security patches and software updates is crucial for protecting your system against vulnerabilities that could be exploited by cybercriminals.

Common System Issues and How to Identify Them

Slow Performance:

Symptoms: Your system takes longer to start, run applications, or complete tasks.

Causes: Overloaded hard drive, too many background processes, outdated hardware, or malware.

Solutions: Free up storage space, close unnecessary applications, upgrade hardware, and run malware scans.

System Crashes and Freezes:

Symptoms: Frequent system crashes or unresponsive applications.

Causes: Overheating, incompatible software, hardware issues, or driver problems.

Solutions: Check for overheating, update drivers, uninstall problematic software, and run hardware diagnostics.

Data Loss:

Symptoms: Missing or corrupted files.

Causes: Accidental deletion, hardware failure, malware, or power outages.

Solutions: Regular backups, file recovery tools, and effective antivirus software.

Internet Connectivity Problems:

Symptoms: Slow or inconsistent internet connection.

Causes: Network issues, ISP problems, or router malfunctions.

Solutions: Restart your router, contact your ISP, or troubleshoot network settings.

Virus and Malware Infections:

Symptoms: Unexpected pop-ups, system instability, or unusual behavior.

Causes: Malicious software infections.

Solutions: Use reputable antivirus software to scan and remove malware.

System Maintenance Best Practices

To prevent common system issues and keep your devices in top shape, consider these best practices:

Regularly update your operating system and software.

Perform routine backups of your data.

Clean your device physically to prevent dust buildup.

Run regular antivirus and malware scans.

Monitor your system's performance and address any issues promptly.

System maintenance is an essential aspect of ensuring the smooth and reliable operation of your devices. By understanding common issues and their causes, as well as following best practices for system maintenance, you can minimize the risk of disruptions and enjoy the benefits of a well-maintained system. Remember that preventative measures and regular attention to your system can save you time, money, and potential data loss in the long run.

Extending the Lifespan of Your Solar Power System

Solar power systems have gained immense popularity in recent years as a sustainable and cost-effective solution for generating clean energy. Whether you've installed solar panels on your home or business, it's essential to understand the importance of regular maintenance and troubleshooting to ensure the longevity and optimal performance of your investment. This guide will provide you with valuable insights into how to extend the lifespan of your solar power system and troubleshoot common issues.

Part 1: Regular Maintenance

Cleaning and Inspecting Solar Panels:

Regularly cleaning your solar panels is a fundamental maintenance task. Dust, dirt, leaves, and other debris can accumulate on the panels, reducing their efficiency. Use a soft brush or a hose to gently clean the surface, being careful not to scratch the glass. Periodic inspections for cracks or damage are also crucial. If you notice any issues, contact a professional for repairs.

Monitoring Solar Inverters:

Solar inverters are critical components that convert the DC power generated by your panels into AC power for use in your home. Regularly monitor your inverters' performance using the provided monitoring system or app. Sudden drops in efficiency or error messages may indicate a problem that needs immediate attention.

Checking Wiring and Connections:

Loose or damaged wiring and connections can lead to power loss and even system failure. Periodically inspect the wiring and connections, ensuring they are secure and free from wear and tear. If you are not comfortable doing this yourself, it's best to hire a professional to perform the checks.

Pruning Nearby Vegetation:

Trees or other vegetation that cast shadows on your solar panels can significantly reduce energy production. Regularly trim or prune such growth to ensure your panels receive maximum sunlight exposure.

Part 2: Troubleshooting Common Issues

Reduced Power Output:

If you notice a decrease in power output, it could be due to shading, dirty panels, or a malfunctioning inverter. Start by cleaning the panels and checking for shading issues. If the problem persists, consult with a solar technician to diagnose and address any more complex issues.

Inverter Errors:

Inverter errors can disrupt your solar power system's operation. Check the inverter display for any error codes or warning lights. Consult the manufacturer's manual to interpret the error and follow the recommended troubleshooting steps. If the issue remains unresolved, contact a professional technician for assistance.

System Not Producing Power:

If your solar power system suddenly stops producing electricity, the problem may be more severe. Check the main electrical panel and circuit breakers to ensure everything is functioning correctly. If there are no apparent issues, contact a professional technician to diagnose the problem.

Monitoring System Alerts:

Many solar power systems come with monitoring systems that send alerts to your phone or email. Pay attention to these alerts, as they can help you detect potential issues early and address them before they become major problems.

Regular maintenance and timely troubleshooting are essential for maximizing the lifespan of your solar power system and ensuring that it operates efficiently. By taking a proactive approach to system care, you can enjoy the benefits of clean and sustainable energy for years to come while minimizing potential downtime and repair costs. If you're unsure about any aspect of maintenance or troubleshooting, don't hesitate to seek professional assistance to keep your solar power system in optimal condition.

CHAPTER 5: LIVING OFF THE GRID

Managing Energy Consumption

Living off the grid is an increasingly popular lifestyle choice that involves disconnecting from the traditional power grid and generating one's energy independently. While this way of life offers many advantages, it also comes with the responsibility of efficiently managing energy consumption. In this article, we'll explore the key strategies for effectively managing energy consumption when living off the grid.

Invest in Energy-Efficient Appliances

One of the most crucial steps in managing energy consumption off the grid is investing in energy-efficient appliances. When selecting appliances for your off-grid home, opt for those with high Energy Star ratings or other certifications that attest to their energy efficiency. Energy-efficient appliances not only reduce your energy consumption but also extend the lifespan of your off-grid energy system.

Prioritize Solar Power

Solar power is a popular choice for off-grid living because it is a clean and sustainable source of energy. Installing solar panels on your property allows you to harness the power of the sun and store it in batteries for later use. By capturing and storing solar energy, you can power your home without relying on traditional utility companies.

Implement a Wind Turbine

In addition to solar power, you can consider implementing a wind turbine if your location allows for it. Wind energy is another renewable resource that can help you generate electricity off the grid. Wind turbines work best in areas with consistent wind patterns, and they can be a valuable addition to your energy production strategy.

Battery Storage Systems

Energy storage is essential when living off the grid. Battery storage systems, such as lithium-ion batteries, store excess energy generated by your solar panels or wind turbine. These batteries allow you to use energy during non-sunlight hours or when there's no wind, ensuring a continuous power supply for your home.

Monitor and Control Energy Usage

Keeping a close eye on your energy consumption is critical for efficient off-grid living. Install energy monitoring systems to track the electricity usage of your appliances and lighting. By identifying energy-intensive devices and habits, you can make informed decisions about when to use them and where to cut back.

Use Energy-Efficient Lighting

Lighting can consume a significant amount of energy in a household. Switch to energy-efficient LED or CFL (compact fluorescent) lighting to reduce your energy consumption. These bulbs use a fraction of the energy of traditional incandescent bulbs and last much longer.

Practice Energy Conservation

Energy conservation is a mindset that can significantly impact your off-grid lifestyle. Simple habits like turning off lights when not in use, unplugging devices, and limiting the use of energy-intensive appliances during peak energy demand can make a substantial difference in your overall energy consumption.

Plan for Seasonal Changes

Living off the grid means adapting to the changing seasons. During the summer, you may have an abundance of solar energy, while the winter months may require more reliance on alternative energy sources. Planning for these seasonal fluctuations can help you manage your energy consumption effectively.

Living off the grid can be a rewarding and sustainable way of life, but it demands a proactive approach to managing energy consumption. By investing in energy-efficient appliances, utilizing renewable energy sources, and adopting smart energy-saving habits, you can enjoy the freedom and self-sufficiency of off-grid living while minimizing your environmental footprint and ensuring a reliable power supply for your home.

Water and Heating Solutions

Living off the grid, a lifestyle choice that involves self-sufficiency and independence from traditional utilities and infrastructure, requires innovative solutions for basic needs like water and heating. In this article, we'll explore water and heating solutions for those who have chosen to embrace off-grid living.

Water Solutions:

Rainwater Harvesting: One of the most sustainable ways to obtain water off the grid is through rainwater harvesting. Collecting rainwater from rooftops into storage tanks can provide a consistent supply of water for various household needs. With proper filtration and treatment systems, rainwater can be made safe for drinking and daily use.

Wells and Boreholes: In areas with access to groundwater, digging a well or borehole can provide a reliable source of water. Hand pumps, solar pumps, or wind-powered pumps can be used to extract water. However, water quality and quantity should be carefully monitored to ensure a continuous supply.

Natural Springs and Streams: For off-gridders located near natural springs or streams, sourcing water from these sources can be an effective solution. Proper filtration and purification methods are essential to ensure the water is safe to drink.

Water Filtration and Purification: Regardless of the water source, an off-grid home should have a robust water filtration and purification system. This can include methods such as UV treatment, reverse osmosis, and gravity filters to remove contaminants and pathogens.

Water Conservation: Off-grid living often involves a conscious effort to conserve water. Low-flow fixtures, composting toilets, and greywater recycling systems can help minimize water usage.

Heating Solutions:

Wood Stoves: Wood-burning stoves are a popular choice for off-grid heating. They provide both warmth and the option to cook food. Sourcing firewood sustainably is crucial to maintain this heating method.

Solar Heating: Solar heating systems can provide a renewable and sustainable source of warmth. Solar water heaters and passive solar design principles can be used to capture and store heat from the sun.

Geothermal Heating: For those living in areas with suitable geological conditions, geothermal heat pumps can be an efficient option. They use the earth's natural heat to warm your home during the winter and cool it in the summer.

Propane and Natural Gas: Propane or natural gas heating systems can be a viable option for off-gridders in some regions. These fuels are stored in tanks and can be used for heating and cooking. However, careful planning is necessary to ensure a consistent supply.

Hydronic Heating: Hydronic heating systems use water or a heat transfer fluid to distribute warmth throughout a home. They can be powered by wood, propane, or other energy sources and are highly efficient.

Living off the grid presents a unique set of challenges and opportunities. To successfully meet the water and heating needs of an off-grid lifestyle, careful consideration of the available resources and appropriate technology is essential. Sustainable and renewable solutions are often favored by off-gridders, aligning with the desire for self-sufficiency and a reduced environmental footprint. Ultimately, the choice of water and heating solutions will depend on the specific circumstances and preferences of those embracing this off-grid way of life

Sustainable Living Off-Grid

In an era dominated by technology and urbanization, the idea of living off the grid has gained significant traction. This lifestyle choice involves disconnecting from the conventional power grid, embracing self-sufficiency, and reducing one's environmental impact. Sustainable living off-grid represents a unique and compelling approach to harmonizing with nature while also promoting a more sustainable future. In this article, we will explore the principles of sustainable off-grid living and its numerous benefits.

Self-Sufficiency:

Living off the grid means relying on your own resources for basic necessities like energy, water, and food. This self-sufficiency is not only empowering but also reduces your dependence on fossil fuels and centralized systems. Solar panels, wind turbines, and hydroelectric generators can provide your energy needs, while rainwater harvesting and well-water systems can secure your water supply. Growing your food or practicing permaculture allows you to become more food independent, reducing your ecological footprint.

Environmental Impact:

One of the central tenets of sustainable off-grid living is minimizing your environmental impact. By producing your energy from renewable sources, you reduce greenhouse gas emissions, air and water pollution, and your contribution to climate change. Furthermore, living off-grid encourages responsible waste management, composting, and recycling, all of which significantly decrease your ecological footprint. Engaging in regenerative practices, such as reforestation and sustainable agriculture, further contributes to a healthier planet.

Reduced Energy Consumption:

Living off the grid often necessitates a more conscientious approach to energy consumption. With limited resources, individuals are encouraged to adopt energy-efficient appliances, reduce energy waste, and implement smart design principles in their homes. This not only conserves resources but also leads to lower utility bills, saving money in the long run.

Connection to Nature:

Off-grid living provides a unique opportunity to reconnect with nature. Living in remote or rural locations allows residents to appreciate the beauty of the natural world, gain a deeper understanding of ecosystems, and foster a stronger sense of responsibility for the environment. People often find that their connection to nature leads to a more fulfilling and peaceful life.

Resilience:

Sustainable off-grid living cultivates resilience in the face of external challenges, such as power outages, natural disasters, or economic instability. By developing skills in food preservation, water conservation, and alternative energy sources, off-gridders can better adapt to unexpected situations. This lifestyle enhances self-reliance, making it easier to cope with adversity.

Community Building:

Living off the grid doesn't necessarily mean isolation. Many off-grid communities are formed by like-minded individuals who share resources, knowledge, and a commitment to sustainable living. These communities create a support system that can enhance the overall quality of life, providing access to diverse skills and experiences.

Sustainable living off the grid offers a unique and rewarding way of life that promotes environmental responsibility, self-sufficiency, and a deep connection to nature. While it may not be suitable for everyone, those who embrace this lifestyle find fulfillment in reducing their environmental impact and fostering resilience in an ever-changing world. As society continues to grapple with environmental challenges and resource limitations, the principles of off-grid living serve as a compelling model for a more sustainable future.

CONCLUSION

The "Off Grid Solar Power Bible: A Step-by-Step Guide to Building Your Own Off-Grid Solar Power System and Go from Zero Know-How to Zero Electricity Bills" is a comprehensive and enlightening resource that has the potential to revolutionize the way we think about energy consumption, sustainability, and self-reliance. Authored by an expert in the field, this book serves as a beacon of knowledge and empowerment for anyone who aspires to liberate themselves from the constraints of conventional electricity grids and embrace a life off the grid.

Throughout the pages of this meticulously crafted guide, readers are taken on a journey that is both educational and inspiring. The author's expertise in solar power systems, combined with a clear and accessible writing style, make the complex world of solar energy accessible to all. Whether you're a seasoned renewable energy enthusiast or a complete novice, the book excels in breaking down technical jargon and simplifying the processes involved in setting up an off-grid solar power system.

One of the book's standout features is its step-by-step approach, which allows readers to proceed at their own pace, gradually gaining the knowledge and confidence required to design, build, and maintain their off-grid solar power system. It instills a sense of achievement as you move through the chapters, from understanding the basics of solar energy to mastering the intricacies of system sizing, component selection, installation, and maintenance. The author's emphasis on safety is commendable, ensuring that readers are well-prepared for the task at hand.

The book goes beyond the technical aspects of off-grid solar power systems, delving into the broader implications of living off the grid. It explores the environmental benefits of renewable energy, highlighting the positive impact on the planet and the reduction of one's carbon footprint. Moreover, it underscores the financial advantages of going off the grid, showing how the initial investment in a solar power system can lead to substantial long-term savings on electricity bills. The book makes a compelling case for how going off the grid can not only provide energy independence but also enhance the quality of life and provide peace of mind in times of power outages or natural disasters.

In addition to practical guidance, the "Off Grid Solar Power Bible" encourages readers to cultivate a deeper understanding of energy sustainability and to take an active role in reducing their ecological impact. It imparts the importance of responsible energy consumption and emphasizes that we all have a part to play in preserving the environment for future generations.

The book also delves into the advancements in solar technology and the evolving energy landscape, ensuring that readers stay up-to-date with the latest innovations and options available for enhancing their off-grid power systems. The wealth of diagrams, illustrations, and case studies further enrich the learning experience, making the content engaging and applicable.

In conclusion, the "Off Grid Solar Power Bible" is more than just a guide; it is an invitation to a brighter, more sustainable future. It provides readers with the tools and knowledge to take control of their energy needs and embrace a lifestyle that is both self-reliant and environmentally responsible. It instills a sense of empowerment and underscores the importance of harnessing renewable energy sources for the benefit of individuals, communities, and the planet at large. As a comprehensive and accessible resource, this book stands as a testament to the potential for positive change when we put the power of the sun to work for us. Whether your goal is to reduce electricity bills, achieve energy independence, or simply to contribute to a greener world, the "Off Grid Solar Power Bible" is an invaluable companion on your journey towards a more sustainable and off-grid future.

Made in United States
Cleveland, OH
31 May 2025

17398232R00066